Everyday NFC
Near Field Communication Explained

Revised and Expanded Third Edition

Hsuan-hua Chang

THIRD EDITION

ISBN-10: 0-9824340-3-0
ISBN-13: 978-0-9824340-3-1

CONTENTS

FIGURES

ACKNOWLEDGMENTS

My appreciation goes to my family who supports me in learning and living my life fully.

David Williams and Julie Chang Schulman are my editors. I thank them for their patience and wisdom.

For this edition I'd like to express additional thanks to all of my readers. Your valued feedback has been deeply appreciated.

Special thanks to:

Koichi Tagawa for providing information on NFC Forum

Randy Vanderhoof for providing information on Smart Card Alliance

Kevin Gillick for information on GlobalPlatform

Poken, ScholarChip, TapWise, Taganize and Tapit for providing information on tag management platforms.

About This Book

Near Field Communication (NFC) is a technology that enables wireless data transfer in close proximity without the need for internet connection. It was approved as an ISO/IEC standard in 2003, and the first NFC-enabled phone was released in 2006. However, the technology has been experiencing slow adoption, especially in the USA.

One product example of its application is demonstrated by a non-profit organization, Khushi Baby. Khushi Baby's mission is to improve vaccination rates in the developing world with an NFC necklace that digitizes data at the point of care. Since 2014, Khushi Baby has been distributing NFC-enabled necklaces to babies in rural India and tracking their vaccination records. This product has successfully facilitated vaccinations for over 10,000 babies.

A less novel, but more recognizable application of NFC is mobile payment. The NFC mobile payment market is largely driven by its rapid adoption in the APAC region and African countries. This adoption is transforming the economics of financial service provision to the poor in developing countries where banking rates are low. NFC is taking mobile payment in bold directions.

NFC technology also helps to enable the Internet of Things (IoT), which is a connected network of physical objects including devices, vehicles, buildings, appliances, electronics, and more. In the world of IoT, objects can exchange information, initiate activities, or activate processes. NFC-enabled IoT was a very hot topic at the 2016 and 2017 Consumer Electronic Show (CES). Oomi, a company that uses a simple tap-to-enable NFC connectivity for a complete home automation system, received an Innovation Award in the Smart Home category at the 2016 CES. At the 2017 CES, more NFC products were introduced that showed how NFC innovates customer service. For example: The "Ocean Medallion" cruise ship uses NFC devices to personalize guests' experiences.

Once consumers and businesses fully adopt this technology, NFC will transform not only the consumer experience, but also our everyday lives.

My goal for writing this book is to raise awareness about the benefits of NFC technology and foster ideas amongst my readers for expanding NFC's applications.

Albert Einstein once said, "If you can't explain it simply, you don't understand it well enough." In this book, I challenge myself to explain NFC technology simply so everyone can understand it with or without a technology background.

Updates:

NFC technology has been evolving quickly since the 2nd edition of this book was published in October 2014. My book is still selling. Whenever a person purchases my book, it encourages me to work more diligently on the latest edition in order to update my readers with developments. You, the reader, are my inspiration. I thank you from the bottom of my heart.

The third edition of "Everyday NFC" covers recent NFC use cases, NFC technical fundamentals, the NFC mobile payment landscape, and its impact on the IoT and Industry 4.0 through the rapid advancements in technology. The information in this edition is new as of November, 2017.

Book Structure:

This book is organized into five chapters:

- **What is NFC**
  A technological overview with examples and summaries.
- **Where is NFC Now**
  Current use cases that demonstrate recent NFC applications in mobile payment, banking, gaming, transit, health care, and more.
- **How to Use NFC**
  Summaries that describe the use of NFC phones/tags/wallet and an overview of the two communication modes, three operating modes, NFC secure transactions and HCE.
- **Who are the NFC Players**
  Summary of the NFC ecosystem, big players, and various industry standard groups.
- **Why Use NFC**
  Summary of the value of NFC and a comparison with other wireless technologies.

I have designed this book for both beginners and more technical readers. The introduction section of each chapter provides a general overview. This is followed by more detailed explanations for a deeper understanding.

Who Should Read This Book?

This book is written for people who are interested in learning about NFC and want to explore the possible uses of this technology. This includes application (app) developers, business leaders, entrepreneurs, innovators, and consumers who are eager to adopt new technology. It is also useful for executives who are responsible for making decisions about NFC projects.

This book is not intended to help develop NFC apps or teach advanced NFC. Therefore, the name of the book is "Everyday NFC".

Supplemental Reading

Go to everyday.com to see the most recent news and discussions.

Go to EverydayNFC at Paper.li to read the daily news about the NFC technology.

Chapter 1: What Is NFC?

Introduction

Near Field Communication (NFC) is a short-range wireless connectivity technology that makes use of interacting electromagnetic radio fields. It has been widely adopted by the public transportation sector, for applications such as the ORCA card in Seattle metro, the SmartTrip card in Washington DC metro, Oyster card in London's tube, Leap card in Ireland and NFC coin in the Taipei Mass Rapid Transit.

When an NFC chip is embedded in a device, the device becomes NFC-enabled. Currently, there are more than 500 million NFC devices on the market (Tagawa, 2016).

NFC-enabled devices can easily connect within four centimeters of each other in order to retrieve and/or exchange data (Figure 1).

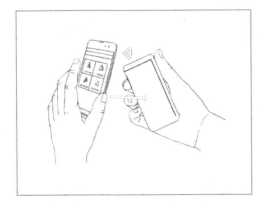

Figure 1: Two NFC-enabled phones exchange data

NFC is based on radio frequency identification (RFID) technology, operates at 13.56 MHz, and transfers data at a rate of up to 424 Kbit/second. Other similar technologies are WIFI and Bluetooth. See Chapter 5 for a comparison of these contactless technologies.

At the same time, NFC is not just a connectivity technology; it is also an enabling technology. For example: NFC can enable mobile payment systems, such as Apple Pay, Google Pay, and Samsung Pay. When you have any NFC mobile payment app installed and activated on your mobile phone, you can forego bringing a wallet with you when you go shopping (Figure 2). See Chapter 3 to learn how to install a mobile wallet and how to use it.

Figure 2: Use NFC phone to card reader for payment

NFC-enabled devices provide additional value when working with NFC tags. NFC tags can be embedded in promotional materials, allowing NFC mobile marketing campaigns to easily reach new audiences. NFC tags can also be used to initiate many automated processes with an NFC-enabled device (Figure 3). For example: NFC can provide home automation of lighting controls or security locks. See the following Technology Overview section for details on NFC tags.

Figure 3: An NFC-enabled phone reads data from an NFC tag

The following are two examples of NFC in action:

1. ORCA Cards

Have you ever ridden a bus in Seattle? In the greater Seattle area, the ORCA card system was launched in 2009 to consolidate payment options for different branches of mass transit. Riders who carry an ORCA card can get on a King County Metro or Sound Transit bus, Light Rail train, or even a ferry without needing cash for the fare. When the ORCA card is swiped across a card reader, the fare is deducted from the card's account value (Figure 4), called an "e-purse" (King County, n.d.).

Figure 4: An ORCA reader reads an ORCA card

Technical summary:

- The ORCA card is a "smart card" with NFC technology.
- A microchip named MIFARE DESfire is embedded inside the ORCA card (Raschke, 2011).
- Information is stored in the microchip, which has a secured microcontroller and internal memory.
- An antenna is embedded in the ORCA card to enable tracking of the card and data transmission.

- When an ORCA card is placed within 4 cm of the card reader, the card reader transfers energy/signal to the microchip in the ORCA card.
  - Wireless communication is established between the ORCA card and the card reader.
  - The data transfer is completed.
  - The ORCA payment process is completed.

2. Assa Abloy's Hospitality Mobile Access

Because NFC operates within such a short range, it is also used for high-security operations such as building access.

For example, when a guest checks in at Prizeotel in Hamburg, Germany, he can access the hotel room by tapping his mobile phone to the lock of the room. Assa Abloy Hospitality Mobile Access enables guests to use their own smart devices as digital room keys in conjunction with VingCard Signature RFID door locks that have also been installed. (Figure 5).

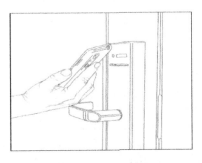

Figure 5: An NFC phone unlocks a door

Technical summary:

- A guest checks in at the hotel.
- The hotel room number is sent to a back-end platform (Assa Abloy) with the guest's information.
- The platform transfers the hotel room number and key over the air to a smart card residing in the guest's mobile phone.
- When the mobile phone is placed within 4 cm of the lock:
  - The wireless communication is established between the phone and the lock.
  - The data transfer is completed between the lock reader and the smartcard in the phone.
  - The room is then unlocked.

Technical Overview

This section provides a technical overview about NFC devices, NFC tags, and NFC Data Exchange Format. Understanding these details can help expand your knowledge on tNFC devices, NFC tags, and how NFC data is exchanged. This is the basic knowledge required to make decisions about investing in NFC technology and developing NFC business applications.

NFC Devices

Currently, there are more than 500 million NFC devices in the market (Tagawa, 2016). IHS forecasted that 2.2 billion NFC cellular handsets will be shipped in 2020 (Tait, 2015) (Figure 6). With increasing shipments year after year, NFC is truly becoming integral to the mobile device.

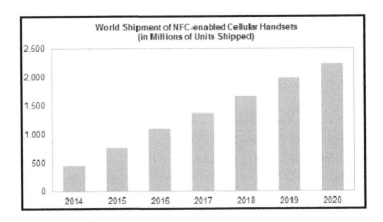

Figure 6: World Shipments of NFC-enabled handsets
Source: IHS Inc. June 2015

- Mobile Phones

    Many smartphones are NFC-enabled. A list of NFC-
    enabled phones can be found at the NFC World web page
    - http://www.nfcworld.com/nfc-phones-list/.

    All NFC smart phones, with the exception of the iPhone,
    can exchange data at a short distance or read/write
    information from/to an NFC tag. NFC-enabled phones
    can also be read by an NFC reader to perform a mobile
    payment. (See Three NFC Operating Modes in Chapter 3.)

    iPhone 6 and subsequent iPhone releases were NFC-
    enabled for Apple Pay that was released in 2016. In
    September 2017, iPhone 7, 7+ and above were enabled to
    read NFC tags via mobile apps with the release of iOS 11.

You might also wonder if you can use your non-smart phone for NFC data exchange. Unfortunately, non-smart phones (so-called feature phones) are not NFC-enabled because they do not contain NFC components.

- Tablets and Laptops

NFC-enabled tablets are a new market segment. More than two dozen brands of NFC-enabled tablets were released from 2012 to 2017 including iPad Pro 10.5-inch released in June 2017, MacBook Pro released in December 2016 , Lenovo ThinkPad X1 Yoga released in August 2016, iPad Pro (9.7-inch) and Samsung Galaxy S7 in March 2016, iPad mini 4 and Asus MeMO Pad 8 in 2015, and the Google Nexus 9, iPad Air2, and iPad mini3 in 2014.

A list of NFC-enabled tablets can be found at the NFC World web page - http://www.nfcworld.com/nfc-data/tablet/.

- Other Devices

Beyond phones, other NFC-enabled devices are also arriving on the market.

The Motiv, a lightweight ring, designed to monitor one's heart rate, track activity, and sleep was announced at the 2017 CES.

Samsung's smart suit, which made its debut in the 2016 CES Conference, can measure heart rate and body fat levels with sensors beneath its fabric (Pachal, 2016).

Apple Watch first generation, first released in 2015,

enables mobile payment for iPhone 5 and later versions. Apple Watch 2, released in 2016, is also NFC-enabled.

On October 1, 2015, almost all credit card readers in US were replaced with EMV compliant readers. EMV (Europay, MasterCard and Visa) is a global standard for credit cards that uses computer chips to authenticate chip-card transactions. Most the EMV-compliant credit card readers are also NFC-enabled.

Tap2Tag introduced NFC wristbands in June 2014 to retrieve a person's medical history (Boden R. , Tap2Tag launches NFC medical alert devices, 2014).

In January 2014, Canon introduced three compact cameras (ELPH 340 HS, N100, SX600 HS) that are NFC-enabled so that they can send photos to and from Android devices with a tap.

Plus Prevention released a range of medical devices. All data from these devices can be transferred by a simple tap to any Android NFC smartphone (Clark S. , 2013) (Figure 7).

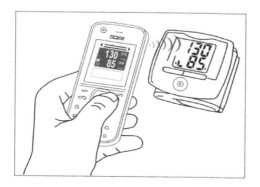

Figure 7: An NFC-enabled phone taps a medical device

NFC Tags

NFC tags (Figure 8) are passive devices used to communicate with active NFC devices. NFC tags can be deployed on physical items which enables connectivity for the Internet of Things.

At the heart of every NFC tag is an NFC chip. It contains a small memory storage chip and a radio chip attached to an antenna. It does not require a power source and can be powered up by an NFC device through a magnetic field (XDA, 2013).

Usually the information stored in the tag's memory is in a specific data format (NDEF) in order to be read by other NFC devices. Please see more details about NDEF in the next section NFC Data Exchange Format.

Figure 8: An NFC tag

The price for a blank tag ranges from US $0.70 to $1.25 based on the size of the memory and its features. Encoding costs range from about US $0.05 to $0.25 per tag based on the content and security requirements.

NFC tags can be programmed and embedded in business cards, smart posters, stickers, wrist bands and promotional materials (Figure 9). They are extremely useful in the distribution of information and the promotion of products and services. They can also launch tasks, preform configurations, and initiate apps when being tapped by an NFC-enabled device.

Figure 9: A user taps his NFC phone on the tag of a poster to retrieve information

Tag Types:

Currently, there are 5 types of NFC tags as follows:

- Type 1:

  - Products examples: Thinfilm OpenSense, Broadcom BCM20203, Innovision Topaz
  - Standards: ISO/IEC-14443A
  - Transfer Speed: 106 Kbps
  - Read or re-write capable; can be configured to read-only
  - Memory availability: 96 bytes and expandable to 2kbytes

- Type 2:

  o Products examples: NXP Mifare Ultralight, NXP Mifare Ultralight C and NTAG21x (with built-in security features)

  o Standards: ISO/IEC-14443A

  o Transfer Speed : 106 Kbps

  o Read and re-write capable; can be configured to read-only

  o Memory availability : 48 bytes and expandable to 2kbytes

- Type 3:

  o Products examples: Sony FeliCa

  o Standards: JIS X 6319-4 (Japanese standard FeliCa)

  o Transfer Speed : 212 or 424 Kbps

  o Read and re-writable, or read-only (pre-configured at manufacture)

  o Memory availability: variable, theoretical limit is 1Mbyte per service

- Type 4:

  o Products examples: NXP DESFire, NXP SmartMX-JCOP, Calypso B

  o Standards: Compatible with ISO/IEC 14443 (A&B)

  o Transfer Speed : 106 Kbps, 212 and 424 Kbps

  o Read, re-writable, or read-only.

- o The memory availability: variable (2kB, 4kB, 8kB), up to 32Kbytes per service

- Type 5

  Type 5 tags are designed to be capable of communicating over much longer ranges (5-10 centimeters) than other tags. It expands the NFC ecosystem to be inclusive of pre-existing professional RFID infrastructure. For example, library books often contain ISO/IEC 15693 tags. An NFC app running on an NFC-enabled device can easily read the book title or ISBN number from the tag.

  - o Products examples: Microelectronics ST25TV
  - o Standards: Compatible with ISO/IEC 15693
  - o Transfer Speed : 6.62 Kbps or 26.48Kbps
  - o The memory availability: variable, the generic memory structure used by Type 5 Tag platform is organized by blocks of fixed size. Each block contains either 4, 8, 16, or 32 bytes. Regular memory offers up to 256 blocks addressed by one byte. Extended memory offers up to 65536 blocks addressed by two bytes.

For Type 5 tag, Operation Technical Specification was released in 2015. Candidate Specification was published in December 2016. These new specifications can be downloaded at the NFC Forum website.

NFC Data Exchange Format (NDEF)

When we tap an NFC-enabled phone to another NFC device or an NFC tag, data is exchanged in the NFC Data Exchange Format (NDEF) that is the standard defined by the NFC Forum. In this section, you will learn about the structure of NDEF and the records it carries.

The NDEF consists of NDEF Messages and NDEF Records. An NDEF Message is an array of NDEF Records with a header and payload (Figure 10). If the payload is large, then the records can be chained to support more data. It depends on the application and tag type to determine exactly how many NDEF Records can be encapsulated in an NDEF Message. For detailed information about NDEF data structure format, please refer to NDEF Technical Specification (Forums, 2006).

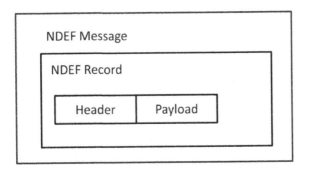

Figure 10: NDEF Message Format

There are four NDEF record types: Text, URI, Smart Poster and Signature. Record Type Definition (RTD) specifies rules for building a standard NDEF Record type for new applications. Signature RTD 2.0 Technical Specification was published in 2015. It allows an NFC device to verify tag data and identify the tag author. When NDEF records are signed in accordance with the Signature RDT 2.0 specification, malicious hackers cannot temper with trusted messages. For detailed information about NDEF record types, please refer to RTD Technical Specification.

Chapter 2: Where Is NFC Now?

Introduction

The global NFC market size was valued at 4.8 billion in 2015 (Research, 2016). The following cart (Figure 11) shows past, current, and forecasted NFC market growth.

Global NFC Market, 2014 - 2024 (USD Billion)

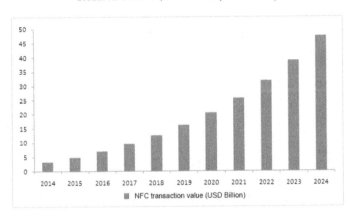

Figure 11: Global NFC Market 2014 – 2024 (USD Billion)

NFC market share has been growing since 2014 because of the following reasons: NFC payment (Apple Pay) enabled by Apple, increasing adoption of the technology into various devices with the demand of IoT, and the relentless effort of ecosystem pioneers.

On the other hand, the growth of the NFC market has been limited by Apple's restricted NFC usage, lack of consumers awareness, and security concerns due to the vulnerability of smartphones.

Device OSs and Software Overview

This section shows how device OSs and software has progressed with NFC.

The following chart (Figure 12) shows that the volume of the NFC-enabled devices shipped has a correlated with the Device OSs development.

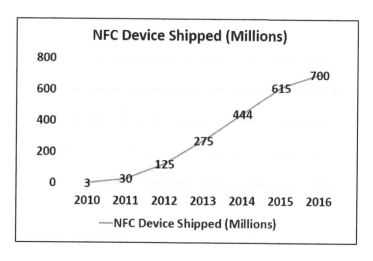

Figure 12: NFC devices shipping stats for 2010-2016

The operating system release timeline (Figure 13) is as follows:

- In 2006, the first NFC-enabled phone (Nokia 6131) was release by Nokia.

- In 2010, the first Android NFC phone (Nexus S) was released by Google.

- In 2011, Blackberry released two NFC phones (Bold 9900 & 9930).

- In 2012, the first Windows Phone (Lumia 610) with NFC was released by Nokia (Davies, 2012).

- In 2014, Apple released NFC-enabled mobile payment iPhone6 with iOS8.

- In 2016, Microsoft released NFC tap to pay with Microsoft wallet for Windows 10 devices, starting with Lumia 950, 950 XL and 650 (Microsoft, 2016).

- In 2016, iPhone7, iPhone 7+, and Apple Watch Series 2 in Japanese markets supported Felica Type-F NFC contactless technology with iOS 10. (Apple, www.apple.com, 2016)

- In 2017, iOS 11 enabled iPhone 7, 7+ and above for reading NFC tags through an app.

- In 2017, iPhone 8, 8+, X and Apple Watch 3 support NFC-F for Felica payments around the world.

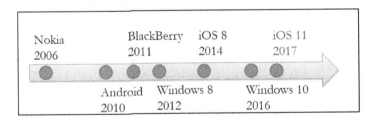

Figure 13: NFC phone release based on OSs

Software Framework

Timelines for software framework releases is as follows:

- In 2010, open NFC API (reader/writer mode) for Android (Gingerbread) was released through NXP and Trusted Logic (News N. , 2010).

- In 2011, Blackberry released Java SDK v7.0 for both tag reading/writing and card emulation apps (Clark S. , 2011).

- In 2011, Peer to Peer mode became available for Android (Admin, 2011). API documentations and NFC sample codes can be found at Android's NFC Basics page.

- In 2013, Windows 8 was released with proximity APIs to support proximity communication using NFC.

- In 2014, MIFARE SDK was released, allowing the development of NFC-enabled Android apps. In September 2016, MIFARE SDK is relaunched under the brand name TapLink.

- In 2016, Windows 10 was released with NFC architecture, a Tap-and-Pay Near-Field proximity provider model, and a universal NFC device driver model.

- In 2017, Apple released iOS11 that enabled NFC tags read through an app. CoreNFC is the framework that supports the use of NFC for reader mode.

- In 2017, NXP releases NFC tag toolkit for iOS 11. For updates, visit: http://nxp.com/nfc.

Internet of Things (IoT)

IoT is a connected network of physical objects including devices, vehicles, buildings, appliances, electronics, and more.

Some of these devices have no user interface.

Gartner predicts that low-power short-range networks will dominate wireless IoT connectivity through 2025 (News G. , 2016), far outnumbering connections using wide-area IoT networks.

NFC provides low-power and short-range connectivity. As an enabler, NFC can interact with these IoT devices for network provisioning, configuration, reporting, and diagnostics with security (Forum, 2016). At the same time, NFC sensors help collect data to enable data analysis in this connected world to improve customer experience, provide early warnings on poor equipment performance, and automate alerts for immediate action or real-time collaboration. (Ransbotham, 2017)

Some examples are as follows:

Connected Car - Honda demonstrated its in-vehicle payment experience in CES 2017 which allows drivers to pay for parking and fuel with a touch of a button on the vehicle dashboard.

Healthcare Environment Monitoring - Monitoring the thousands of portable cylinders used to refill larger tanks at medical facilities across many states used to be labor intensive. Nowadays, inexpensive sensors embedded in each cylinder can monitor contents, capture fill-level and pressure data, and send critical information over wireless networks to data centers that facilitate delivery truck dispatch.

Industry 4.0

Industry 4.0 was an initiative started by the German government in 2006. It has also been called the "industry internet" or "the fourth industrial revolution." The initiative's intention is to digitize the manufacturing sector in order to increase productivity.

There are nine key technological components that make up the foundation of Industry 4.0: autonomous robots, big data, augmented reality (AR), additive manufacturing, cloud computing, cyber security, IoT, system integration, and simulation.

NFC, which adds new levels of convenience, communication, and configurability throughout the manufacturing process, also collects and exchanges data to provide insights and help improve processes.

A good example is how Smart Factory uses sensors to enable automation. For example a Scalance XM-400 basic device from Siemens is an integrated NFC solution for initiating mobile diagnostics (Siemens, 2015).

NFC tags are used to verify the authenticity of individual components to ensure that robots use the right item for a given task (Humber, 2015).

Mobile Wallet

A mobile wallet leverages NFC technology to enable contactless payments at point-of-sale terminals using NFC-enabled smartphones (See Mobile Wallet history in Figure 14). The amount of money spent around the world using a mobile wallet will rise by nearly 32% this year to US$1.35tn, Juniper Research Predicts. (Boden R. , 2017)

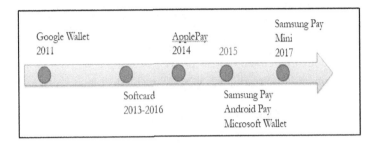

Figure 14: Mobile Wallet History

An NFC mobile wallet is composed of a few components: a mobile application (app), payment options (credit cards), an authentication method for user identification, and an NFC chip for wireless transmission or Host Card Emulation (HCE) software structure for an app to emulate a card and talk to an NFC card reader directly.

During the set-up of a mobile wallet app, your credit card information can be entered and stored into your mobile wallet. At the check stand, you open the app on your phone, tap or hold it against the cashier card reader, and your payment will be processed with the selected credit card.

Currently, there are a few mobile payment apps as follows:

Apple Pay is a NFC mobile wallet app provided by Apple launched with iOS 8.1 in October 2014 in US with international roll-out afterwards. Now, Apple Pay is available in over 20 countries It works with the iPhone 6/6+ and sequential releases, iPad Air2, iPad Pro, iPad Mini3/4 and even the iWatch when paired with an iPhone 5 and sequential iPhone releases. An NFC chip is embedded in the Apple devices to enable the Apple Pay tapping and paying.

Samsung Pay was released in August 2015 in Korea and 2016 in US. Not only does it work with NFC, Samsung phones are also equipped with Mobile Secure Transmission (MST) technology that emulates a swipe transaction through the swipe reader in case the reader is not NFC-enabled.

Samsung also released Samsung Pay Mini in 2017. Samsung Pay mini can run on all Android phones that are running Android 5.0 Lollipop and above as long as they have a screen resolution of 1280x720 or higher. It enables mobile payment on Samsung's affordable devices that don't qualify for the full-fledged Samsung Pay service functionality.

Android Pay was released in September 2015. It works with all NFC-enabled Android devices running KitKat 4.4 and above and Android Watch with Android Wear 2.0 and above. Android Pay can also be used on iOS devices through Android Wear 2.0. Android Pay uses Host Card Emulation (HCE) technology to interact with NFC payment terminals.

Microsoft Pay is available in Microsoft insider. Windows 10 for phones support Host Card Emulation (HCE). HCE allows any smartphone with Windows 10 and NFC hardware to transmit payments from the device to an NFC terminal designed to receive that money without needing a special secure app or any secure SIMs from wireless carriers.

NFC-ready Point-Of-Sale (POS) is used for contactless

payment with NFC-enabled mobile devices. It is another indicator for the growth of NFC mobile payment.

Berg Insight's research has revealed that the global installed base of NFC-ready POS terminals will grow at a compound annual growth rate of 28.4% from 21.4 million in 2014 to 74.9 million in 2019. See the growth chart below (Figure 15).

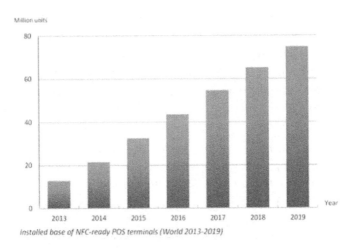

Installed base of NFC-ready POS terminals (World 2013-2019)

Figure 15: Installed base of NFC-ready POS terminals

More Use Cases & Current NFC Products and Services Examples

In recent years, early adopters have started focusing on the creation of NFC products and services. NFC applications continue to be released.

Below are examples of applications which demonstrate the versatility of NFC technology.

Wearable

Wearables are still at their infancy stage, however NFC-enabled wearable devices will likely be at the center of future market growth. Other than smartwatches and wristbands, NFC tags can be embedded in clothing, shoes, and accessories.

- A clever use of an NFC tag has been integrated into Google's Daydream View VR headset and its do-it-yourself VR kits, Cardboard. Mounting your NFC-capable smartphone into the headset triggers the nearby tag to automatically download or launch the app (Faulkner, 2017).

- Moscow metro commuters can uses an NFC ring (PayRing) to pay their fare anywhere that accepts the Troika transport card (Moscow Metro to Introduce Fare Payment Rings, 2017).

- Motiv, a lightweight ring, designed to monitor one's heart rate as well as track activity and sleep was announced at CES 2017 and shipped in September 2017. It is also waterproof down to 50 meters and can handle being separated from your phone for up to five days at a time for record synchronization (Hartmans, 2017).

- Arrow's latest range of Smart Shirts are enabled with NFC chips. They can be paired with Android smartphones and interact with the Arrow mobile app to do various tasks; for example configuring your phone to play your favorite music (Mathur, 2016).

Gaming

NFC earned a role in gaming a few years ago. In 2013, McDonald's announced the rollout of Happy Table in Asia. This video (http://bit.ly/1euNXQB) shows how NFC tags are utilized to transform a regular table into a gaming platform.

More recent products are as follows:

- FusionPlay – Heros was the first mobile NFC card game released in 2017. It contains an NFC-chip in every playing card. This enables the cards to communicate with a smartphone app that makes the cards come alive on the screen of a phone. It merges together physical items with digital applications to create an unique gaming experience. (Kunze, 2017)

- Lego Dimensions launched on the Xbox One, Xbox 360, PS4, PS3, and Wii U in 2015. It is a toys-to-life crossover video game where a player has Lego figures and a toy pad that can be played within the game itself. It features characters and environments from over 30 different franchises. The Starter Pack, containing the game, the USB toy pad, and three mini-figures were released in September 2015, and has since been expanded with additional level packs and characters (Becker, 2017).

- Nintendo has been leveraging NFC to enrich the gaming experience of Wii U players. For example, players can now use the NFC interface built into the Game Pad Controller to purchase games with a prepaid e-card.

Nintendo has also integrated NFC into gaming itself. Amiibo is an NFC-enabled figurine that can gain skills and attributes proportional to user interaction. When a player touches Amiibo to the Game Pad, his character's data is downloaded into the game, and he can also send information back to his character. Therefore, Amiibo's attributes get a small boost in each battle which increases overall level, defense, and attack powers in a game. (June 14) (Boden R. , Nintendo unveils Amiibo NFC figures that work with multiple Wii U and 3DS games, 2014)

In addition to increasing purchasing and playing power, NFC is also allowing Nintendo users to import photos and content from their phones to their game consoles through a peer-to-peer data exchange mode (wiiudaily, 2014).

Transportation:

- Swedish train operator SJ Railways has become the latest company to adopt technology that reads NFC microchips embedded in the body as an alternative to printed rail passes. There are now around 3,000 commuters using the technology. (Morris, 2017)

- Huawei worked with NXP to roll out NFC mobile payments Huawei Pay for public transport in Shanghai, Shenzhen, Guangzhou, and Beijing in 2016 ((Boden R. , Huawei Pay NFC mobile payments rolled out across China public transport, 2016)

- London Metro started to use NFC cards in 2012 and accepted NFC-enabled phones for payment in 2014.

In 2013, London Metro started to offer news at 10,000 bus stops in the UK via NFC. By using NFC devices to tap tags at any Metro touch point ad Panel, people can connect to the Metro landing page in order to access news. This is an example of using NFC tags to direct consumers to a designated website.

HealthCare

- GlaxoSmithKline NFC smart shelves can be scanned by NFC-enabled phones to receive medical product information on the shelves. GlaxoSmithKline plans to use Thinfilm's software platform CNECT to deliver customer messages, view real-time consumer messages, view real-time consumer activity, and view other analytics built into the platform((Muoio, 2017).

- Micro Instruments has already worked with innovative companies on optimal selection, placement, and processing of semiconductor strain gages for implantable NFC sensors for the heart, the brain, the spine, and specialized post-surgical monitoring (Chelner, 2016)

- Tag2Tag is leading the field in medical wearables producing a medical alert device that is designed to help first-responders and paramedics gain instant access to critical medical information. Tag2Tag stores users' medical information on a secured server and allows NFC-enabled cards, wristbands, and NFC tags to access the information in an emergency situation. In this case, NFC is used as an access device (Tap2Tag, 2017)

Consumer Electronics

- NFC-enabled cameras can transfer photos easily with another NFC-enabled devices such as phones or printers with a tap. Nikon, Canon, Panasonic, and Sony all have released NFC-enabled cameras((Beren, 2017).

- The DuraScan® Model D600 Contactless Reader/Writer is the future of the data capture industry. It makes the transitioning from barcode scanners to NFC readers simple. It features great ergonomics, fitting perfectly in your hand for comfortable and extended use (Mobile, 2017).

- NFC Touch to print was enabled a few years ago. By downloading an app into an NFC-enabled phone and tapping the phone onto the Samsung Xpress printer, one can easily print, scan, or fax in 2013. Nowadays, most of the major printer makers—HP, Brother, Canon, Epson have implemented NFC on many of their inkjet and laser printers to allow touch-and-print functionality ((Harrel, 2016)

- Sony unveiled a family of photographic zoom lenses with integrated image sensors that can be connected to Android or iOS smartphones via NFC and WiFi to provide a high quality, DSLR-like camera experience. This is an example of using NFC to extend the functionality of smartphones

(Clark M. , Sony launches NFC lens-style cameras for smartphones, 2013)

Event Managements.

- The Oakland Athletics let fans enter their stadium by tapping their phone (or Apple Watch) to a ticket scanner via Apple Pay. This was the first time a professional sports team supported contactless tickets in Apple Wallet (Tepper, An MLB team is using the iPhone's NFC feature for contactless stadium entry, 2017)

- Poken supported NFC utilization at the 2014 Youth Olympic Games in Nanjing. Each participant in the event received a Yogger, an NFC-enabled device which was used to exchange contact and social media information with other participants (BusinessWire, Nanjing Youth Olympic Village Promotes High-Tech Socializing, 2014)

In 2017, Poken was acquired by GES, a global, full-service partner for live events.

Smart Label and Packaging

- Thinfilm prints NFC chips onto plastics and enables easy communication between brands and consumers through the packaging. For example, the smart label that Thinfilm prints and puts on a bottle of Johnnie Walker Blue Label scotch is a way to authenticates Diageo's bottle, showing it's not counterfeit (Thinfilm's tiny printed NFC labels enable brands to connect with consumers via smart packaging, 2017)

- Thinfilm introduced NFC smart labels with temperature sensors that can be used for food packaging or tracking wine temperature. Consumers can use an NFC-enabled phone to check whether a product has been kept at its required temperature through the supply chain (Boden R. , Thinfilm shows printed NFC smart label with temperature sensor, 2014).

Insights from Data Collection

- Thinfilm launched its CNECT software portal in 2017. It's a multi-tenant cloud-based platform integrated with NFC SpeedTap and OpenSense NFC tags. It provides a simple and secure way to store, manage and track the tags to provide consumer insights.

Marketing

- In 2014, to celebrate the expansion of the Polo range at Harrods, 15 window displays at their store in London were brought to life with NFC-enabled touch points. Passers-by could tap their phones to the windows and access exclusive content, download a map of Harrods, and even order and buy online if the store was closed (Top 5 Contactless Campaigns using RFID & NFC, 2015).

- Crosscliq published Canada's first Blackberry NFC-enabled magazine ad. NFC tags were embedded in the Rogers Connected Magazine. When an ad is tapped by a Blackberry NFC-enabled phone, a holiday gift is offered. This is an example of embedding NFC tags into printed material.

- Tenet Computer Group Inc. released the GreenRack Cloud service that enables people to globally distribute information. Tapping your smartphone to a GreenRack station launches the Information e-Concierge, which automatically scans associated QR codes or NFC tags. This is an example of combining location-based content with on-demand publishing.

Chapter 3: How To Use NFC On Your Smartphone

Introduction

It is exciting to test NFC functionalities when you obtain an NFC-enabled phone. You can scan a tag or tap another NFC-enabled phone to exchange data, or you can explore your mobile payment options. This section shows you how.

Using NFC with an Android or Windows phone

- **To activate NFC functionality:**
    1. Go to Home.
    2. Go to Applications (or Apps on Android phone only).
    3. Go to Settings.
    4. For Android: check the NFC app
       For Windows 8 and above: check the tap+send app.

- **To read an NFC tag:**
    1. You do not need to download an app in order to read or scan an NFC tag.
    2. Place your phone over the tag within 4 cm (Figure 16).
    3. When asked, accept the request to receive the data exchange.
    4. If it doesn't work, make sure that the NFC functionality is enabled and the screen is active.

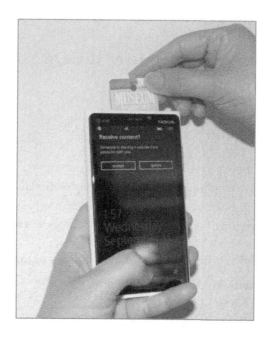

Figure 16: Tap an NFC-enabled phone to read an NFC tag

- **To Write to an NFC Tag using an Android Phone**
  1. Make sure that the NFC functionality is enabled on the device.
  2. Download NXP TagWrite app from Google Play Store.
  3. Click on "Create, write and store" to create a NFC data set.
  4. Type in a URL to a web page.
  5. Tap your phone to an NFC Forum Type 2 Tag when you see the screen display: "Ready to store or share the selected content" as shown (Figure 17).

Figure 17: TapWriter app is ready to write content to an NFC Tag

6. When you see "Store successful" message on the Result window, this confirms that new content has been written to the tag.

Using Android Pay:

NFC is available for Android Pay with Android 4.4 (KitKat) and above.

1. Make sure that the NFC functionality is enabled on the device.
2. Look for the Android Pay app on your phone. It is preloaded on several devices. You can also simply download it from the Google Play Store.

3. Add a credit or debit card into the app. If you already have a card in your Google account, you can add it to Android Pay.

4. Use the NFC device to pay your bill at stores that have NFC readers.
   1) Tap your phone near the contactless reader
   2) Default credit card will be charged.

5. You can also use an Android Wear Watch if it has an NFC chip and Android Wear 2.0. Simply press the bottom button of the watch to access the feature.

Using NFC with an iPhone

Using Apple Pay

NFC functionality is limited to Apple Pay in iPhone6, 6s and newer iPhone models.

- **To Use Apple Pay Mobile Wallet**
  1. Make sure you have iOS 8.1 or above (Settings/General/Software Updates shows iOS version)
  2. If you have a supported credit or debit card on file with iTunes, enter the card's security code. Otherwise add a new card to Wallet by doing the following steps:

     Click on Settings/Wallet & Apple Pay/Add Credit or Debit Card to add a new card to Wallet.

     Use your iPhone camera to capture and store your card number, then answer other card related questions in the next step, or you can enter the card information manually.

3. If you are wearing an Apple Watch that is paired with your iPhone, you will be provided an option to use "Add Now" to add Apple Pay to your Apple Watch in the next step.
4. The first card you add automatically becomes your default card.
5. Use iPhone to pay your bill at stores that have NFC readers:
   1) With your finger on Touch ID, hold your iPhone near the contactless reader.
   2) Make the payment with your default credit card or choose another credit card to pay by selecting a new default in Settings. You can go to Wallet at any time to pay with a different card.

- **To Use Apple Watch for Apple Pay**
  1. Pair your iPhone 5, 5s, 6, 6s, SE, and above with iWatch
     1) Click on the Watch app on your iPhone.
     2) Start pairing as directed by the iPhone (Figure 18)

  2. Pay with Your Apple Watch
     1) Double-click the side upper button

2) Choose Wallet icon; then card to use
3) Double-click the side lower button and hold the display of your Apple Watch within a few centimetres of the reader until you feel gentle pulse.

Figure 18: iPhone pairs with Apple Watch

Technology Overview

This section explains the technology behind NFC functionalities. It includes NFC communication modes, NFC operating modes, Hosted Card Emulation (HCE), and the secure mode components that ensure the security of mobile transactions. This information will give you additional background for NFC operation and infrastructure.

Two NFC Communication Modes

The two communication modes are active and passive.

- **Active mode**:
  Two NFC devices can tap each other for data
  exchange with their own power supply. (Figure 19).

Figure 19: NFC-enabled phones exchange data in
active mode

- **Passive mode**:
  The primary NFC-enabled device generates radio
  signals, and the receiving device is powered by the
  magnetic field of the primary device. This mode allows
  a passive NFC tag to be read by an active NFC reader.

  An ORCA Card transaction (Figure 20) is an example
  of passive mode usage. The ORCA card itself is always
  in passive mode and does not have a power supply.

The ORCA card reader is in active mode. When the ORCA card is placed close to the reader, the reader powers the ORCA card and starts the data exchange.

Figure 20: A passive NFC card is read by an active NFC reader

Three NFC Operating Modes

NFC-enabled devices support three modes of operation; Reader/Writer, Peer-to-Peer, and Card Emulation

- **Reader/Writer**: This is an open/non-secure mode; an NFC-enabled device acts like a reader or writer to read from or write to an NFC tag or another NFC device that operates in the card emulation mode. The mode is based on the ISO 14443 standard and FeliCa schemes. Android phones can read an NFC tag without a mobile app, but requires a mobile app to write to an NFC tag. iPhone 7 and above need a mobile app to read NFC tags with iOS 11.

- **Peer-to-Peer (P2P)**: This is an open/non-secure mode; two NFC-enabled devices can exchange data based on the ISO 18092 standard, which includes two modes; P2P passive and P2P active. Either device can initiate communications.

An example of this is the exchange of digital business cards (Figure 21).

In this case:

o After introductions, one party initiates the exchange with an NFC-enabled device and sends a business card to the other party.

o The other party sends a business card back to the initiator.

Two NFC standards are used in this case:

o Logical Link Control Protocol (LLCP): The NFC Forum has specified this protocol to enhance the Peer-to-Peer mode of operation. LLCP sits on top of the ISO 18092 in the protocol stack and is used to establish communication.

Simple NDEF Exchange Protocol (SNEP): This is an application-level protocol released by the NFC Forum for sending or receiving messages between two NFC-enabled devices.

Active Communication Mode (ACM) was introduced to support P2P communication in 2015 and adopted by NFC technical specifications in 2016.

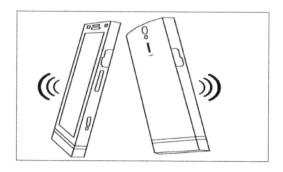

Figure 21: Digital business cards are exchanged when two NFC-enabled phones tap each other

- **Card Emulation**: This is a secure mode; an NFC device is used in passive mode to emulate the behavior of a contactless card which is based on the ISO 14443 standard. Mobile payment is implemented in this mode (Figure 22).

  In a mobile payment transaction:

  o An NFC reader generates a magnetic field through its antenna.
  o The reader sends a command to the NFC-enabled phone the same way it would send a command to a contactless smart card.
  o The NFC-enabled phone sends its response to the NFC reader.

Figure 22: An example of card emulation

Beyond the user interface layer, in a mobile NFC system, the service provider and the mobile network operator interact via their respective NFC platforms to ensure management and administration of the mobile NFC services using the NFC operating modes of Reader/Writer and Card Emulation (Figure 23).

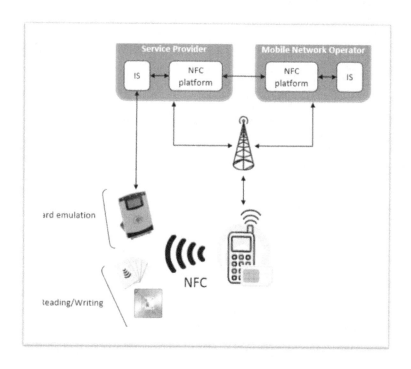

Figure 23: NFC system overview from AFSCM

NFC Secure Transaction Secure Element (SE) Approach

An NFC secure transaction supports secure card emulation services such as mobile wallet and building access. Understanding the secure transaction processes provides important background for designing or choosing a secured NFC product or service.

Using SE for an NFC Secure transaction, an infrastructure might include a Secure Element (SE), Over-the-Air (OTA) download via a Trusted Service Manager (TSM) and a Trusted Execution Environment (TEE) in some cases (Figure 24).

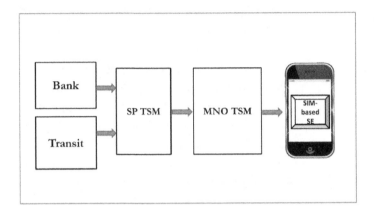

Figure 24: Infrastructure of a SE secure transaction

- **Secure Element (SE)**

  The Secure Element (SE) is a tamper resistant smart card that includes secure microcontrollers, a CPU, an operating system, memory, and a crypto engine. It is used to store sensitive information for NFC products and services. For example, a hotel room access application can store hotel key information on the SE or credential information to authenticate mobile payment.

  The current implementation of SE has the following formats:

- o Hardware Format:
  - ➢ Embedded: eSE in the phone, this is a device manufacturer centric approach. Apple uses this approach. Therefore, iPhones come with a Universal Integrated Circuit Card (UICC) that has a Subscriber Identification Module (SIM) to identify subscribers.
  - ➢ Removable: UICC/SIM-based SE, this is the telecom operator centric approach. Softcard used to use this approach. The carriers have control over storage and access of the SE (Figure 25).
  - ➢ Attached: MicroSD in a phone sleeve or tag, this approach is used for Apple devices that don't have NFC capability.

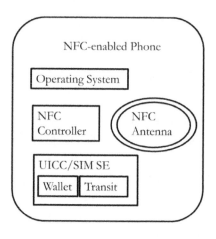

Figure 25: NFC-enabled phone with UICC/SIM SE

o Software Format:

> Cloud-based: SE is stored in the Cloud. Google Wallet uses this approach. When a consumer conducts a transaction, encrypted data will be pulled from the SE in the Cloud (Figure 26).

> Not Cloud-based: SE is stored in the phone application.

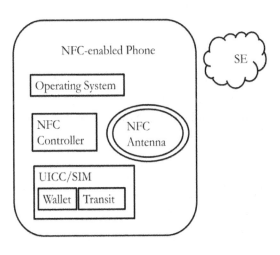

Figure 26: Cloud-based SE

- **Trusted Service Manager (TSM)**

  Storing, accessing and managing information in a removable SIM-based SE requires an infrastructure that facilitates over-the-air (OTA) download and management of SE. One approach, endorsed by the GSMA, is to deploy

TSM; a connection point between service providers, such as banks and transit operators, and SE. TSM must be compliant with Global Platform standards and GP messaging specifications (Global Platform, 2013).

There are two types of TSM as follows:

o **Mobile Network Operator (MNO) TSM**
MNO TSM performs management of credentials and OTA provisioning stored in the SIM-based SE that is in the mobile device on behalf of the service provider. It performs distribution, provision and management of the NFC apps in the SE (See Figure 22).

MNO TSM may have the following capabilities:
➢ Interconnection with Mobile Network Operations (MNO) and Service Providers (SP)
➢ MNO management
➢ SP management
➢ SP application management
➢ OTA provisioning and mobile device management
➢ Maintenance of end-to-end security

o **Service Provider (SP) TSM**
SP TSM acts as a bridge between service providers and MNO TSM that operate by mobile network operators to enable mobile commerce such as mobile payments (See Figure 22).

SP TSM may have the following capabilities:
➢ Management of multiple MNO and SP interfaces
➢ Management of rule engines and workflow

- ➤ Management of applets and security domains (SDs) on SE
- ➤ Lifecycle management of NFC service applications
- ➤ OTA support
- ➤ OTA card management for all SE implementations
- ➤ Customer service
- ➤ Reporting capabilities
- ➤ Billing support
- ➤ Direct communication with the SE for card management functions without connecting through the MNO TSM

NFC secure transaction using SE approach has been adopted by Softcard Mobile Wallet. It's a telecom operator-centric approach since telecoms own TSMs (the access of the SIM-based SE) and the real estate on the SIM. Softcard Mobile Wallet, which was launched in November 2013 and discontinued in March 2015, used this approach. Right now, US mobile payment vendors are not using this approach, but it can be used for other secure transaction applications in the future.

Because of this trend, TSM vendors have started to move away from the telecom centric approach to NFC applications. For example, Giesecke & Devrient (G&D) is working with chip maker Renesas to apply TSM technology to protect industrial products against counterfeiting. (Clark M. , G&D uses TSM technology to protect industrial products against counterfeiting, 2016)

NFC Secure Transaction Host Card Emulation (HCE) Approach

HCE was a term created by the founders of SimplyTapp, Doug Yeager and Ted Fidelski, in 2011. It describes the ability to transact with remotely operated smart cards.

HCE's card emulation allows applications to emulate a card and talk to an NFC reader directly. There is no need to have a SE in this case. Google Wallet was re-architected from a SIM-based SE to an HCE model to remove dependency on telecom operators. HCE was incorporated into the Android 4.4 operation system to support NFC secure transactions which include payments, loyalty programs, card access, and transit passes.

Security against authorized account access in HCE relies on four key concepts: limited use keys, tokenization, device fingerprinting, and dynamic risk analysis as follows:

- Limited use keys (LUK) are derived from a master domain key shared by the issuer and the cloud card management vendor.
- Tokenization reduces risk for banks by replacing the Personal Account Numbers (PAN) with a tokenized pseudo-PAN used in the payment system.
- Device profiles or "fingerprints" are intended to ensure transactions are initiated only by authorized user devices.
- User, device and account data is used to perform risk assessment for the transaction in real time (Roy, What is Hosted Card Emulation, 2014)

SE vs. HCE approach

- SE approach: in NFC card emulation mode, the card to be emulated is provisioned into a SE on the device through an app. When the user holds the device over an NFC reader, the NFC controller in the device routes data from the reader to the SE where the credential is saved for authentication (Figure 27).

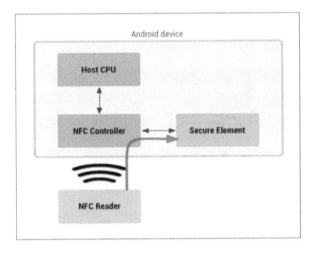

Figure 27: NFC card emulation with a secure element
Source: developer.android.com

- HCE approach:
  When the user holds the device over an NFC reader, the NFC controller in the device routes data from the reader to the host CPU on which applications are running (Figure 28).

HCE simplified implementation is being adopted in various NFC secure applications including banking. The cost efficiency of HCE allows for more secure application development. This approach is helping to increase NFC adoption.

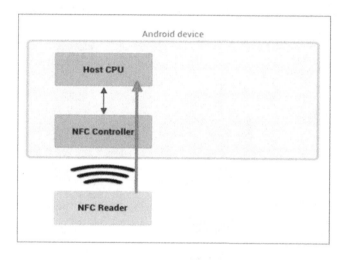

Figure 28: NFC card emulation without a SIM-based SE
Source: develop.google.com

Trusted Execution Environment (TEE) for Mobile Security

The TEE is a secure area that resides in the main processor of a mobile device. It allows applications to execute, process, protect, and store sensitive information in its trusted environment (Figure 29). There are different architectural ways to achieve TEE. The Global Platform works to standardize specifications for TEE and standardization enhances mobile security, and NFC secure mode operations that use UICC/SIM-based SE. The specifications for card, device and systems can be found at www.globalplatform.org/specifications.asp.

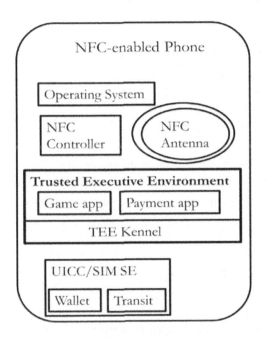

Figure 29: Trusted Execution Environment

In August 2015, Global Platform published a System Protocol Discovery Mechanism (SPDM) specification that eased NFC roaming by enabling an NFC system to discover the application protocols that are supported by a second NFC system.

In January 2017, Global Platform published the TEE management framework that details how trusted applications hosted on a Global Platform-compliant TEE can be remotely and dynamically managed.

In July 2017, Sierraware announced that it made Global Platform-compliant TEE available for MIPS-based devices to meet the security demand of the applications for the connected embedded devices.

Arm Trustzone Security Approach

Trustzone is Arm's hardware isolation technology that was released in 2003. It evolves over time based on the security requirements of mobile devices. It is a System on Chip (SoC) and CPU system-wide approach to security.

Trustzone enables a single physical processor core to execute from both the Normal world and the Secure world. Normal world components cannot access secure world resources (Figure 30).

In order to implement a secure world in the SoC, trusted software (trusted OS) is developed to make use of the protected assets. This software typically implements trusted boot, the secure world monitor, a small trusted OS, and trusted apps. The combination of Trustzone, trusted boot, and a trusted OS makes up a TEE.

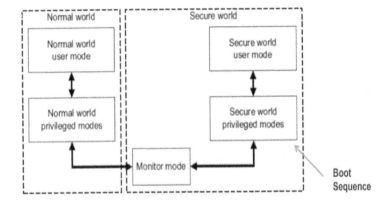

Figure 30: Arm trustzone

Management of Multiple Contactless Card Emulation Environment (CEEs)

The management of CEEs hosted in the same device is not standardized. The Global Platform has proposed the Managing Entity specification to define a framework that provides a whole standardized ecosystem offering a reliable and interoperable environment for Service Providers willing to deploy mobile contactless service.

In addition, the Global Platform Managing Entity specification defines a fallback mechanism to manage legacy Secure Element (e.g. UICC or eSM).

Chapter 4: Who Are The NFC Players?

Introduction

The NFC Ecosystem has been growing during the past few years. The unique combination of mobile and NFC technology has expanded the ecosystem.

Ecosystem players include manufacturers that produce mobile devices, smart cards, tags, readers and consumer electronics. Other ecosystem players are semiconductor manufacturers, service providers including telecom operators, financial institutions, system integrators, certification organizations, merchants, marketing vendors, the printing and publishing industry and app developers.

When a new technology emerges, stakeholders often join efforts to form standards groups to advance the technology and minimize proprietary implementation.

Standards Groups

Currently, there are standards groups working on general NFC technology, as well as standards groups working on NFC devices. The standards landscape interactive PowerPoint presentation is a good tool to understand the standards used in different NFC products and services.

The following standards groups are significant players in the NFC ecosystem. A brief description of the groups and their contribution towards the NFC standards are as follows:

NFC Forum

NFC Forum was established in 2004 and leads the effort for the unification of the NFC ecosystem. More than 140 companies are members of the NFC Forum. They share their expertise to develop the specifications and provide approval and certification for products. More than 20 specifications have been released (Figure 31). For more details about the specifications, please visit NFC Forum website at http://www.nfc-forum.org.

Apple joined the NFC Forum in August 2015 as a sponsor member and joined the board of directors.

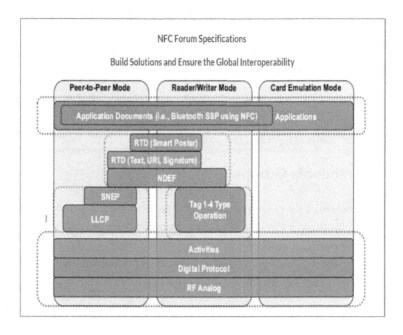

Figure 31: NFC Forum Specification Architecture

Source: NFC Forum goo.gl/IuUEmc

The NFC forum certification program was launched in 2010. It provides device manufactures with a means of establishing that their products conform to the NFC Forum specifications. In 2017, NFC tags and readers are added into the certification program.

NFC Forum has also established a brand symbol "N-Mark" (Figure 32) as the universal symbol and touch point for NFC. It shows consumers where to touch their NFC-enabled device to initiate NFC actions. The guidelines for proper usage of N-Mark are provided at http://nfc-forum.org/our-work/nfc-branding/n-mark/

Figure 32: N-Mark

Secure Technology Alliance (Formerly Smart Card Alliance)

Formed in 2001, Smart Card Alliance was a multi-industry association working to promote the understanding, adoption, use and widespread application of smart card technology. In 2017, with over 200 members, Smart Card Alliance changed its name to Secure Technology Alliance due to its expanded charter. Now its activities also cover embedded chip technology and related hardware/software that support the

implementation of secure solutions. See more details at
https://www.securetechalliance.org/.

Global Platform

Launched in 1999, Global Platform has over 100 members. It
standardizes the management of applications that are on the
secure chip. It defines specifications for cards, devices and
systems, as well as streamlining security requirements and
testing through security certifications. These specifications and
certifications are highly regarded as the international standard
for enabling digital services and devices to be trusted and
securely managed throughout their lifecycle.

GlobalPlatform has accredited 14 labs in China, France,
Germany, Korea, Netherlands, Spain, U.K., and U.S. and made
18 test suites available. See more details at
http://globalplatform.org

SIMalliance

SIMalliance, founded in 2000, is a global, non-profit industry
association which simplifies secure element (SE)
implementation to drive the creation, deployment and
management of secure mobile services. See more details at
www.simalliance.org

The World Wide Web Consortium (W3C)

W3C, founded in 1994, is the main international standards
organization for the World Wide Web. It defines an Open

Web platform for application developers to build rich interactive experiences. It published its first working draft on Web NFC API in 2014 and an update was released in 2015. See more details at www.w3.org/TR/nfc

NFC Device Standards

Many groups are involved in the standardization of NFC-enabled devices. These efforts include integrating NFC technology into mobile devices, making sure that NFC transactions are secure and interoperable in the architecture design.

- **GSM Association (GSMA)**

    Formed in 1995, the GSMA is an association of mobile operators and related companies devoted to supporting the standardization, deployment and promotion of the GSM mobile telephone system. See http://www.gsma.com/.

- **International Organization for Standardization /International Electrotechnical Commission (ISO/IEC)**

    Founded in 1947, the ISO/IEC publishes International Standards covering almost all aspects of technology and business.

    For NFC standards, ISO/IEC focuses on the card emulation interfaces. See http://www.iso.org.

- **European Computer Manufacturers Association (ECMA) International**

  Founded in 1961, ECMA is dedicated to the standardization of Information and Communication Technology (ICT) and Consumer Electronics (CE). (ECMAWeb, n.d.)

  ECMA focuses on the interface between an NFC transceiver and a contactless front end (CLS). It works closely with ISO/IEC. See http://www.ecma-international.org/.

- **The European Telecommunications Standards Institute (ETSI)**

  Created in 1988, ETSI produces standards for Information and Communications Technologies (ICT), including fixed, mobile, radio, converged, broadcast and internet technologies (ETSIWeb, n.d.)

  ETSI focuses on interface between NFC APIs and the NFC controller. See http://www.etsi.org.

- **Java Community Process (JCP)**

  Introduced in 1998, JCP is the open, participative process to develop and revise the Java technology specifications, reference implementations and test suites (JCPWeb, n.d.)

  It focuses on JSR 257 and 177. See http://jcp.org.

- **Open Mobile Alliance (OMA)**

  Formed in June 2002, OMA delivers open specification for creating interoperable services that work across all geographical boundaries (OMAWeb, n.d.).

  OMA focuses on browser and SCWS in the USIM for the UMTS network. See http://www.oenmobilealliance.org.

- **3rd Generation Partnership Project (3GPP)**

  Formed in 1998, 3GPP provides a stable environment to six telecommunications standard development organizations (ARIB, ATIS, CCSA, ETSI, TTA, TTC) in order to produce specifications that define 3GPP technologies (3GPPWeb, n.d.).

  3GPP focuses on NFC modem standardization. See http://www.3gpp.org.

- **Europay, MasterCard, Visa Cooperation (EMVCo)**

  Formed in 1993, EMVCo is a global standard for credit and debit payment cards based on chip card technology. EMVCo chip based payment cards are smart cards that contain an embedded microprocessor that has information needed for payment. (EMVWeb, n.d.)

In March 2014, the EMVCo published Payment Tokenization Specification (EMVCo, 2014) set the framework for tokenizing contactless payment.

In 2016, EMVCo released Level 1 specification that defined the physical characteristics, radio frequency interface and transmission protocol between credit and debit cards and payment terminal.

In 2017, EMVco released Payment Tokenization Specification – Technical Framework V2.0 to introduce new roles and include ecommerce use cases and operational management enhancements to support global interoperability and facilitate transaction security. See http://www.emvco.com.

Chip manufacturers

There are many integrations occurring with NFC in chip technology.

For example, NXP launched two new NFC chips in 2017 as follows:

- NTAG 413 DNA enables a secure one-time authentication code to be generated when the tag is tapped by an NFC device.

- NTAG 213 Tag has a tamper-evident feature and can be placed on the product label, seal or container so that product information can be accessed with a tap of an NFC device.

Both NXP chips are designed to advance product authentication, integrity assurance and enhance user engagement across consumer manufactured goods, healthcare, retail and other industries in an IoT world.

On October 24th 2017, NXP announced the new LPC8N04 MCU, an integration of NFC and LPC800 series of microcontroller. This innovation disrupted microcontroller landscape and ushers a new wave of consumer and industrial IoT application (GlobeNewswire, 2017).

Another example is newly released ST53G system-in-package solution from STMicroelectronics. It combines an enhanced NFC radio with a secure banking chip in one compact module to enable secure payment to wearables.

ST offers an extensive development ecosystem, including radio-tuning tools and pre-defined antenna configurations. The ST53G meets all relevant card-industry standards, including EMVCO compliance, ISO/IEC-14443 NFC card emulation, and MIFARE ticketing specifications.

Other than NXP Semiconductors (Netherlands) and STMicroelectronics (Switzerland), the other key chip manufactures' players are as follows:

- Broadcom Corporation (U.S.)

- Intel Corporation (U.S.)
- Qualcomm Inc. (U.S.)
- Texas Instruments (U.S.)
- Samsung Semiconductors (South Korea)
- Austria Micro Systems (Austria)
- ST Micro (Switzerland)
- MStar Semi (Taiwan)
- Flomio (U.S.)
- Gemalto (Netherlands)

Chapter 5: Why Use NFC?

Introduction

"NFC creates a world of secure universal commerce and connectivity in which consumers can access and pay for physical or digital services, anywhere at any time with any devices." This was the vision shared by the NFC Forum at the Wireless CTIA in 2009, and it still rings true for technology visionaries across the globe.

ABI Research supported this vision when it reported that, "NFC is not just in mobile phones, tablets, PCs and peripherals, speaker docks, televisions, cameras, gaming and domestic appliances are all increasingly incorporating NFC" (ABI Research, NFC Installed Base, 2013).

Utilizing NFC can add the following values to your business:

- **Simplicity**: NFC simplifies data capture and expedites information collection, transfer and distribution. Unlike other connected technology, NFC doesn't require an app to exchange data or download information.
- **Accessibility**: NFC enables consumers to access content and services with a tap. This provides convenience for the consumers, especially those of the Y generation that are very connected to their mobile phones.
- **Security**: SIM card-based NFC provides ultimate security for mobile payment or non-payment secure applications such as transit or building access.

- **Cost efficiency**: NFC SIM cards and tags are relatively inexpensive compared to other currently utilized enablers.
- **Versatility**: NFC brings value to other electronics and creates new business opportunities.
- **Energy efficiency**: The passive nature of NFC chips doesn't require power resource.

Technology Overview on RFID, QR Code and Bluetooth

When people start understanding NFC technology, they often want to know what the differences are among NFC, QR Code and Bluetooth. Since NFC is based on RFID, questions are also asked about the differences between NFC and RFID. This section reviews these wireless connectivity technologies for curious minds.

NFC vs. RFID

NFC is based on RFID technology. RFID stands for Radio Frequency Identification. It is a small electronic device that consists of a chip and an antenna.

Here are some key points of contrast between NFC and RFID:

- NFC
  - Limited range of communication (4 centimeters)
  - Two-way communication
  - Three operating modes which enable a wide range of applications

- RFID
  - Tags are either active or passive. Active RFID tags can broadcast with a read range of up to 100 meters. Passive RFID tags have a read range of up to 25 meters (RFIDInsider, 2013).
  - One way communication: i.e. a reader detects and pulls information from a tag.
  - When more than one reader overlaps, an RFID tag is unable to respond to simultaneous queries. This is called Reader Collision.
  - When many tags are displayed in a small area, readers will be confused by all these signals. This is called Tag Collision.

NFC vs. QR code

A QR code (Figure 33) is a two dimensional bar code that can store information. QR stands for Quick Response. It was invented in 1994 in Japan for the automotive industry and started to gain market share when its adoption rate began rising.

Figure 33: QR code

ScanLife released a survey entitled "Mobile Barcode Trend Report 2013 Q3" (Figure 34) which revealed some interesting statistics:

- Scanning activity is dramatically on the rise, now surpassing 180 scans per minute.
- There has been a 77% increase in new users.
- Android OS is responsible for 64% of QR code scans while iOS is responsible for 34%.

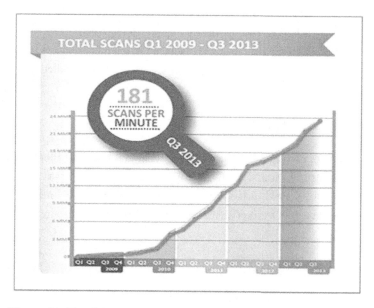

Figure 34: Total Scans Q1 2009 – Q3 2013 (LifeScan)

Q1 2014 survey showed that Mobile engagement is rapidly increasing, totally at 21.8 million scans this quarter. This activity is a 20 % increase from last year. iOS scan is up 46% from last year.

Here are some key points of contrast between NFC and QR codes:

- NFC
  - Needs NFC-enabled devices but does not require applications to use.
  - Not based on images scanned, but rather embedded tags that do not require visual cues to be detected by a device.
  - Tags are better for branding since they can be printed with color.
  - Tags are rewritable.
  - Tags have a secure mode option.
  - Communication mode is active to active or active to passive.

- QR code
  - Can be read by all smartphones with the proper applications.
  - Costs less than NFC.
  - Have market recognition.
  - Communication mode is passive; waiting for readers to scan.

NFC vs. Bluetooth

Bluetooth and NFC are both wireless connectivity technologies that communicate between devices over short distances.

Bluetooth was invented by Ericsson in 1994. It can connect 8 devices simultaneously and overcome problems of synchronization. When Bluetooth-capable devices come within range of one another, an electronic communication takes place to determine whether they have data to share, or whether one needs to control the other. Once the communication is complete, the devices form a network (Layton).

The differences between NFC and Bluetooth are as follows:

- NFC
  - Slower transfer speed (424 kbit/s)
  - Frequency 13.56 MHz
  - Communication mode is active to active or active to passive
  - Shorter range of connectivity (within 4 centimeters)
  - NFC's passive devices do not consume power
  - Works with passive devices

- Bluetooth
  - Faster speed (721 kbit/s)
  - Frequency 2.4 GHz
  - Longer range of connectivity (32 feet)
  - Communication mode is active to active
  - Conventional Bluetooth consumes more power, although the newer Bluetooth low energy (BLE) consumes less power than NFC
  - Does not work with passive devices
  - Requires devices to be paired by manual setup prior to communication
  - Data transfer is highly encrypted

BLE and Beacons

Bluetooth Low Energy (BLE)/Bluetooth Smart was originally introduced in 2006. It was merged into the main Bluetooth standard in 2010 with the adoption of the Bluetooth Core Specification Version 4.0.

BLE features:

- Ultra-low peak, average and idle mode power consumption
- Ability to run for years on standard coin-cell batteries
- Low cost
- Multi-vendor interoperability
- Enhanced range

With BLE transmission, a data packet is received by a device, a beacon ID is extracted from the packet by the OS, and the ID is made available to the appropriate app. The app then interrogates the ID to determine the next action (Stemle, 2014).

NFC vs. BLE

An NFC device transmission is one-to-one. It can't be used to send push notifications.

BLE-enabled devices receive signals from Beacons (wireless transmitters). Since a BLE signal is always broadcasting, it can perform one-to-many transmission.

NFC and UHF

UHF (Ultra High Frequency) is a type of passive RFID transponder that can be read at several meters however it operates at much higher frequencies than NFC.

UHF is used in applications such as library books, and luggage counting on conveyor belts. The tags are extremely low cost however the readers are more expensive than NFC readers.

When an application requires hundreds of thousands of tags strictly for identification and only a small number of readers, UHF can be a good option. In applications where there are many readers, NFC can be more cost effective.

Summary and Looking Ahead

Summary

This book provides an overview of NFC technology by answering the following frequently asked questions:

- What is NFC
- Where is NFC Now
- How to use NFC
- Who are the NFC Players
- Why Use NFC

After reading through the book, you now have a basic understanding of NFC technology and development worldwide.

As a business leader, it is important to include an NFC strategy as a part of your business strategy; especially in proximity marketing and smart packaging areas while NFC applications are emerging.

As an app developer, recent use cases listed in chapter 2 can help you envision products and services that can tap into this technology. You might want to explore what apps will be in high demand in the near future.

As a consumer, your knowledge can make an impact on the development of the technology and adoption for other consumers.

Starting in 2018, New York's Metropolitan Transportation Authority will begin moving to a contactless payment system that uses mobile apps like Apple Pay, Android Pay, or NFC contactless credit cards (Abrams, 2017). However, London public transportation already adopted contactless card and mobile payments in 2014. Therefore, I want to reflect what challenges causes this 3-years delay.

The following retrospective serves as lessons learned.

Retrospective

The slow adoption rate that has limited the growth of NFC technology in the US is due to two main factors. One is the lack of general NFC awareness. The other is slow adoption from Apple.

Lack of NFC general awareness

Most of consumers who have NFC-enabled phones don't know about NFC applications. Samsung has attempted to address this issue through its Galaxy advertisements; however, more work in the area of education needs to occur in order to increase user awareness.

This point became evident in the case of Softcard Mobile Wallet, where the consumer awareness impacted the adoption rate. Softcard was rolled out nationwide in November 2013 by AT&T, T-Mobile and Verizon as a joint-venture. However as the first NFC mobile payment service in the USA, the adoption rate was not high enough to sustain the service. Softcard service was discontinued in March 2016.

The following case also demonstrates an impact caused by insufficient NFC awareness in the banking industry: banks were mandated to issue EMV compliant chip cards in late 2016. However since they opted to save costs by issuing chip cards without contactless antennas, the opportunity to launch contactless usage in over 3 million full EMV compliant and contactless capable terminals was tragically missed.

Compared to popular tap and pay transactions in Europe, Asia, and Canada, "insert a chip card" payment method has left the US behind on the contactless payment trend.

Slow adoption of NFC by Apple

The following statistics show the impact that Apple could have made if it had adopted NFC earlier. In 2013, 275 million Android and Windows NFC enabled devices were shipped. At the same time, global iPhone sales was over 150 million. However, iPhone was not NFC capable. Retailers and other solution providers were hesitant to go into this marketplace without NFC-enabled iPhone.

In 2013, not only did Apple failed to adopt the NFC technology, but it also embraced BLE with iPhone 5s/5c release. It was disappointing for NFC enthusiasts and BLE has been falsely framed as a competing technology to NFC.

In 2014, Apple finally provided an NFC solution for its mobile wallet with the release of Apple Pay for iPhone6/6+. It was very encouraging for the NFC ecosystem. However, other than Apple Pay, no other NFC functionalities have not been activated.

In September 2017 with the release of iOS 11, iPhone 7 and above can read NFC tags via mobile apps. It's exciting news since Apple finally is providing the software framework for NFC reader mode. Brand activation, product authentication, product insights, and user guides can all be launched with a simple tap. This will have quite an impact given Apple serves more than 40% of the US Smartphone market (Lovejoy, 2017). So what's ahead?

Looking Ahead

NFC will start ramping up more rapidly for the following reasons: security, data collection, Apple's coreNFC expansion, new technologies on tags/chips, blockchain, and ease of use.

Security

Since consumers are depending more on their mobile devices for a variety of services, mobile security is becoming a central concern. The fact that NFC could store sensitive information in a Secure Element is very appealing to consumers. The following table (Figure 35) shows the growing shipments of worldwide Secure Elements.

Worldwide Secure Elements shipments & forecasts: 2016-2017
(Millions of units)
Source: Eurosmart, May 2017

WW shipments forecast	2015	2016	2017f	2016 vs 2015 % growth	2017f vs 2016 % growth
Telecom	5,300*	5,450*	5,550	3%	2%
Financial services	2,850	2,900	3,000	2%	3%
Government - Healthcare	410	460	510	12%	11%
Device manufacturers	310	330	390	7%	18%
Others**	450	470	490	4%	4%
Total	9,320	9,610	9,940	3%	3%

*Source SIMalliance
**Others include Transport, PayTV and physical and logical access cards

Figure 35: Worldwide SE shipment & forecast 2016-2017

Insights based on NFC solutions

For businesses, NFC solutions bring a wealth of information about consumer buying habits, product and service consumption rates, connected devices, and the effectiveness of marketing campaigns. NFC moves big data into a new era.

New NFC products are being introduced at a more rapid pace with the increasing availability of NFC-enabled devices, standardizations and technology awareness. Innovators also started to use P2P mode to create a better mobile wallet experience. For example, Bank of American announced that in the 3rd quarter of 2017, the total number of peer-to-peer payments made over the Zelle network by its customer base jumped 68% on a year-over-year basis to 13.6 Million, bringing total payment volume up to 40% to $4 Billion.

Expansion of Apple's CoreNFC

When it comes to iOS apps, coreNFC, Apple's new software framework is a promising start. NFC reader apps are able to read NFC tags with the new framework. Apple will continue to expand coreNFC's functionalities to writer mode and P2P. When that occurs, iPhones will become a full NFC enabled device. That will really help promote NFC technology. It is clear that Apple is using NFC as a commodity for consumers to upgrade to newer iPhone models.

Given Apple's inclusion of NFC in its Apple Pay, and now NFC reader offer, we can anticipate a widespread adoption of NFC by Apple users in the near future.

New technologies drive down the cost of tags

New technologies are producing lower cost NFC tags and more integrated NFC chips, and this trend will help move the technology forward. For example, in February 2017, Cartamundi, Imec, Holst Centre and TNO unveiled plastic NFC tags. That offers the ability to manufacture chips in large volumes at low cost.

Blockchain increases the speed of tracking

A blockchain is a tool of recording, maintaining and discovering information on a distributed ledger with minimal third party involvement. When tracking NFC tags with blockchain, data collection will be very fast.

Ease of use

Tapping is quite an easy motion. In Canada, UK and Australia , the use of NFC contactless cards are skyrocketing. Tap-and-pay attracts consumers for its ease of use and speed. When US banks start to issue NFC contactless cards, tap-and-pay will be adopted quickly and based on trend occurring in Canada, this adoption rate will be faster than NFC mobile payment.

In summation, I am envisioning a world where NFC is part of our everyday lives. More products and services will come to the market when consumers and businesses see the value of using the technology. Consumers will become familiar with tapping in order to retrieve information and exchange data; as well as making an appointment, paying bills, monitoring health, transit use and accessing buildings.

Hopefully this book has sparked your imagination about the possibilities of NFC in your life and business. Be a technology pioneer and take advantage of this growing field. Thank you for reading.

Glossary

Acronym	Definition
AFSCM	Association Française du Sans Contact Mobile
API	Application Programming Interface
ARIB	Association of Radio Industries and Businesses, Japan
ATIS	Alliance for Telecommunications Industry Solutions, USA
BLE	Bluetooth Low Energy
CAGR	Compound Annual Growth Rate
CCSA	China Communications Standards Association
CLF	Contactless Frontend
Contact Smart Card	Require contact to initiate a transaction
DSLR	Digital single-lens reflex cameras
ECMA	European Computer Manufacturers Association

Acronym	Definition
EMV	Europay, MasterCard and Visa
EMVCo	Europay, MasterCard, Visa Cooperation
ETSI	European Telecommunications Standards Institute
iOS	iPhone OS (Operating System)
G&D	Giesecke & Devrient
GSMA	Global System for Mobile Communications Association
HDMI	High-Definition Multimedia Interface
IC	Integrated Circuit
IS	Information System
JCP	Java Community Process
ISO/IEC	International Organization for Standardization/ International Electrotechnical Commission
LLCP	Logical Link Control Protocol (LLCP)
MCU	Microcontrol Unit

Acronym	Definition
MNO	Mobile Network Operator
MWC	Mobile World Congress
NDEF	NFC Data Exchange Format
NFC	Near Field Communication
OMA	Open Mobile Alliance
ORCA	One Region Card for All
OTA	Over the Air
POS	Point of Sale
QR	Quick Response
RTD	Record Type Definition
RFID	Radio Frequency Identification
SCWS	Smart Card Web Server
SE	Secure Element
SIM	Subscriber Identification Module
SNEP	Simple NDEF Exchange Protocol
SP	Service Provider

Acronym	Definition
TEE	Trusted Execution Environment
TSM	Trust Service
TTA	Telecommunications Technology Association, Korea
TTC	Telecommunication Technology Committee, Japan
UICC	Universal Integrated Circuit Card
UMTS	The Universal Mobile Telecommunications System
USIM	The SIM application for UMTS network
URI	Unified Resource Identifier
USIM	Universal Identity Subscriber Module
USB	Universal Serial Bus
W3C	The World Wide Web Consortium
WIFI	Is a technology that allows an electronic device to exchange data or connect to the internet wirelessly using radio waves

Acronym	Definition
Y Generation	There are no precise dates for when Y Generation starts and ends. Commentators use beginning birth dates from the early 1980 to the early 2000s.

Works Cited

3GPPWeb. (n.d.). *About 3GPP*. Retrieved 9 14, 2013, from http://www.3gpp.org: http://www.3gpp.org/About-3GPP

ABI Research. (2012, 11 2). *NFC will Come Out of the Trial Phase*. Retrieved from ABIresearch: http://www.abiresearch.com/press/nfc-will-come-out-of-the-trial-phase-in-2013-as-28

ABI Research. (2013, 3 26). *NFC Installed Base*. Retrieved from ABI Research: https://www.abiresearch.com/press/nfc-installed-base-to-exceed-500m-devices-within-1

ABLOY, A. (2013, March 4th). *Seos : Powering Mobile Access*. Retrieved from ASSA ABLOY: http://www.assaabloy.com/en/com/Press-News/News/2013/Seos-rides-NFC-wave-at-Mobile-World-Congress/

Abrams, A. (2017, 10 23). *The New York Subway Will Phase Out MetroCards In Favor of Apple Pay*. Retrieved from Fortune: http://fortune.com/2017/10/23/new-york-subway-metrocards-apple-pay/

Admin, N. (2011, dec 28). *Development with Android Beam and NFC Peer-2-Peer*. Retrieved from www.nfc.cc: http://www.nfc.cc/2011/12/28/development-android-beam-and-nfc-peer-2-peer/

Allprnews. (2013, 8 13). *Near Field Communication Applications Market* . Retrieved from http://allprnews.com/: http://allprnews.com/near-field-communication-applications-market-to-reach-10015-96-million-by-2016-at-a-cagr-of-38-from-2011-to-2016-new-report-by-marketsandmarkets

Apple. (2014, 9 12). *Apple Pay.* Retrieved from www.apple.com: http://www.apple.com/iphone-6/apple-pay/

Apple. (2016, 9 7). *www.apple.com.* Retrieved from Newsroom: http://www.apple.com/newsroom/2016/09/apple-pay-coming-to-japan-with-iphone-7.html

Balaban, D. (2013, 7 31). *Qualcomm to Integrate Trusted Execution Environment with NFC Technology.* Retrieved from NFC Times: http://nfctimes.com/news/qualcomm-integrate-tee-nfc-technology

Becker, D. (2017, 3 21). *The Evolution Of Dimensions Toy Tags.* Retrieved from brickstolife: http://www.brickstolife.com/the-evolution-of-dimensions-toy-tags/

Beren, D. (2017, 9 15). *The 8 Best NFC Cameras to Buy in 2017* . Retrieved from Lifewire: https://www.lifewire.com/best-nfc-cameras-493511

BluetoothSIG. (2014). *Bluetooth Smart (Low Energy).* Retrieved from Bluetooth Developer Portal: https://developer.bluetooth.org/TechnologyOvervie w/Pages/BLE.aspx

Boden, R. (2013, 8 13). *NFC World.* Retrieved from Coca-Cola

runs NFC promotion in 100 stores:
http://www.nfcworld.com/2013/08/13/325454/coca
-cola-runs-nfc-promotion-in-stores/

Boden, R. (2014, 6 11). *Nintendo unveils Amiibo NFC figures that work with multiple Wii U and 3DS games*. Retrieved from MFC World:
http://www.nfcworld.com/2014/06/11/329632/nint
endo-unveils-amiibo-nfc-figures-work-multiple-wii-u-
3ds-games/

Boden, R. (2014, 6 19). *Tap2Tag launches NFC medical alert devices*. Retrieved from http://www.nfcworld.com:
http://www.nfcworld.com/2014/06/19/329805/tap2
tag-launches-nfc-medical-alert-devices/

Boden, R. (2014, 5 28). *Thinfilm shows printed NFC smart label with temperature sensor*. Retrieved from NFCWorld:
http://www.nfcworld.com/2014/05/28/329390/thin
film-shows-printed-nfc-smart-label-temperature-
sensor/

Boden, R. (2015, June 30). *NFC World*. Retrieved from NFC World:
http://www.nfcworld.com/2015/06/30/336312/2-
2bn-nfc-enabled-handset-shipments-by-2020-ihs-
technology-predicts/

Boden, R. (2016, 12 20). *Huawei Pay NFC mobile payments rolled out across China public transport*. Retrieved from NFCWorld:
https://www.nfcworld.com/2016/12/20/349129/nfc

-mobile-payments-rolled-across-china-public-
transport/

Boden, R. (2016, 12 20). *Huawei Pay NFC mobile payments rolled
out across China public transport.* Retrieved from
NFCWorld:
https://www.nfcworld.com/2016/12/20/349129/nfc
-mobile-payments-rolled-across-china-public-
transport/

Boden, R. (2017, 2 2). *Mobile wallet spending to rise 32% in 2017.*
Retrieved from www.nfcworld.com:
https://www.nfcworld.com/2017/02/02/349810/mo
bile-wallet-spending-rise-32-2017/

Boden, R. (2017, 2 8). *World first' plastic NFC tag opens up new
possibilities for NFC deployments.* Retrieved from NFC
World:
https://www.nfcworld.com/2017/02/08/349995/49
995/

Buckley, S. (2013, 9 5). *Verizon Wireless shows off NFC wares
during the NY State Fair.* Retrieved from
www.fiercetelecom.com:
http://www.fiercetelecom.com/story/verizon-
wireless-shows-nfc-wares-during-ny-state-fair/2013-
09-05

BusinessWire. (2013, 7 30). *Isis® Announces National Rollout
Later This Year.* Retrieved from
www.businesswire.com:
http://www.businesswire.com/news/home/20130730
006909/en/Isis%C2%AE-Announces-National-
Rollout-Year

BusinessWire. (2014, 8 22). *Nanjing Youth Olympic Village Promotes High-Tech Socializing*. Retrieved from Business Wire: http://www.businesswire.com/news/home/20140822 005137/en/Nanjing-Youth-Olympic-Village-Promotes-High-Tech-Socializing#%2EU_xig7y1Yc8

Cattaneo, P. (2014). TEE Conference . *Global Platform Second Annaul Conference* (pp. Identity, authentication & Payments Panel). Santa Clara: Global Platform.

Chang, H.-h. (2013, 9 7). *Apple and NFC*. Retrieved from EverydayNFC: http://everydaynfc.com/2013/09/07/apple-and-nfc/

Chelner, H. (2016, 10 12). *Micro Instruments*. Retrieved from https://www.microninstruments.com: https://www.microninstruments.com/news.aspx?sho warticle=14

Clark, M. (2013, 9 4). *Sony launches NFC lens-style cameras for smartphones*. Retrieved from NFCworld: https://www.nfcworld.com/2013/09/04/325724/son y-launches-nfc-lens-style-cameras-smartphones/

Clark, M. (2016, 1 29). *G&D uses TSM technology to protect industrial products against counterfeiting*. Retrieved from https://www.nfcworld.com/: https://www.nfcworld.com/2016/01/29/341782/gd-uses-tsm-technology-to-protect-industrial-products-against-counterfeiting/

Clark, S. (2011, 5 31). *RIM releases BlackBerry NFC APIs*.

Retrieved from NFC World:
http://www.nfcworld.com/2011/05/31/37778/rim-releases-blackberry-nfc-apis/

Clark, S. (2013, June 12). *ABI reports NFC chip market shares.*
Retrieved from NFO World:
http://www.nfcworld.com/2013/06/12/324581/abi-reports-nfc-chip-market-shares/

Clark, S. (2013, 8 13). *TapCheck to launch NFC medical devices.*
Retrieved from NFC World:
http://www.nfcworld.com/2013/08/13/325472/tapc
heck-to-launch-nfc-medical-devices/

Clark0925, S. (2012, 9 25). *Inside Secure to offer cloud-based NFC
secure element solution.* Retrieved from
http://www.nfcworld.com/:
http://www.nfcworld.com/2012/09/25/318059/insi
de-secure-to-offer-cloud-based-nfc-secure-element-solution/

Davies, J. (2012, 4 11). *Hands on: The Lumia 610, Nokia's first
Windows NFC phone.* Retrieved from NFC World:
http://www.nfcworld.com/2012/04/11/315025/han
ds-on-the-lumia-610-nokias-first-windows-nfc-phone/

ECMAWeb. (n.d.). *What is ECMA.* Retrieved 9 14, 2013, from
http://www.ecma-international.org:
http://www.ecma-international.org/memento/index.html

EMVCo. (2014, 3). *Payment Tokenisation.* Retrieved from
www.emvco.com:
http://www.emvco.com/specifications.aspx?id=263

EMVWeb. (n.d.). *About EMV*. Retrieved 9 13, 2013, from
https://www.emvco.com/:
https://www.emvco.com/

ETSIWeb. (n.d.). *About ETSI*. Retrieved 9 14, 2013, from
http://www.etsi.org: http://www.etsi.org/about

Faulkner, C. (2017, 5 9). *What is NFC? Everything you need to
know*. Retrieved from Techradar:
http://www.techradar.com/news/what-is-nfc/2

Forum, N. (2016, 6 29). *Simplifying IoT: Connecting, Commissioning,
and Controlling with Near Field Communication – NFC
Makes the Smart Home a Reality*. Retrieved from
http://nfc-forum.org: http://nfc-
forum.org/newsroom/definitive-internet-things-nfc-
white-paper-published-nfc-forum/

Forums, N. (2006, 7 24). The NFC Data Exchang Format
technical specification.

Francisco Corella, P. (2014, 9 25). *Smart Cards, TEEs and
Derived Credentials*. Retrieved from Promcor:
http://pomcor.com/2014/09/25/smart-cards-tees-
and-derived-credentials/

Gillick, K. (n.d.). *GlobalPlatform made simple guide: Trusted
Execution Environment (TEE) Guide*. Retrieved 9 12,
2013, from http://www.globalplatform.org:
http://www.globalplatform.org/mediaguidetee.asp

Global Platform. (2013, 2 27). *GlobalPlatform Specification
Supports Service Provider Integration into the NFC Ecosystem*.

Retrieved from http://www.globalplatform.org/:
http://www.globalplatform.org/mediapressview.asp?i
d=982

GlobeNewswire. (2017, 10 24). *NXP Integrates NFC Technology
into LPC800 Series Microcontrollers Revolutionizing Smart
Tagging in IoT Applications*. Retrieved from NASDAQ:
http://www.nasdaq.com/press-release/nxp-
integrates-nfc-technology-into-lpc800-series-
microcontrollers-revolutionizing-smart-tagging-in-
20171024-01126

Harrel, W. (2016, 10 17). *Near-Field Communication (NFC),
Mobile Device Printing* . Retrieved from lifewire:
https://www.lifewire.com/print-with-your-mobile-
device-2769176

Hartmans, A. (2017, 9 24). *This $200 ring is the first activity tracker
I actually want to wear all day, every day.* Retrieved from
businessinsider:
http://www.businessinsider.com/motiv-ring-activity-
tracker-review-photos-2017-9/#one-key-feature-sets-
it-apart-on-board-memory-2

Heisler, Y. (2014, 10 2). *Apple Pay: An in-depth look at what's
behind the secure payment system*. Retrieved from
www.tuaw.com:
http://www.tuaw.com/2014/10/02/apple-pay-an-in-
depth-look-at-whats-behind-the-secure-payment/

http://paybefore.com. (2014, May 15th).
http://paybefore.com/pay-news. Retrieved from
http://paybefore.com: http://paybefore.com/pay-
news/isis-touts-600000-monthly-account-activations-

as-pace-doubles-may-15-2014/

http://www.fastcasual.com. (2014, 7 23). *http://www.fastcasual.com/news*. Retrieved from http://www.fastcasual.com/news/jamba-juice-gives-away-millionth-beverage-to-isis-mobile-wallet-user/: http://www.fastcasual.com/news/jamba-juice-gives-away-millionth-beverage-to-isis-mobile-wallet-user/

Hubmer, P. (2015). *What NFC Means for Smart Factories, Intelligent Supply Chains and Industry 4.0*. N.V.: NXP.

Humber, P. (2015). *What NFC means to Smart Factories, Intelligent Supply Chains, and Industry 4.0*. N.V.: NXP.

Hunter, P. (2014, 9 10). *NFC Forum*. Retrieved from www.nfc-forum.org: http://nfc-forum.org/newsroom/analysts-apple-pay-will-help-spur-wider-usage-of-mobile-payments/

IHS. (2014, 2 27). *NFC-Enabled Cellphone Shipments to Soar Fourfold in Next Five Years* . Retrieved from http://press.ihs.com: http://press.ihs.com/press-release/design-supply-chain/nfc-enabled-cellphone-shipments-soar-fourfold-next-five-years

JCPWeb. (n.d.). *FAQ*. Retrieved 9 14, 2013, from http://jcp.org: http://jcp.org/en/introduction/faq#speclead

Khoo, N. A. (2012, 8 23). *You can finally use NFC for payments in Singapore*. Retrieved from Cnet: http://asia.cnet.com/you-can-finally-use-nfc-for-payments-in-singapore-62218405.htm

King County. (n.d.). *Fares & ORCA*. Retrieved 9 11, 2013,
   from http://metro.kingcounty.gov:
   http://metro.kingcounty.gov/fares/orca/index.html

Kunze, K. (2017, 5 17). *www.mifare.net*. Retrieved from A new
   gaming genre – NFC card games :
   https://www.mifare.net/mistory/a-new-gaming-
   genre-nfc-card-games/

Layton, C. F. (n.d.). *How Bluetooth Works*. Retrieved from
   Howstuffworks:
   http://electronics.howstuffworks.com/bluetooth.htm

Li, C. (2016, 2 26). *Is Near Field Communication (NFC) finally
   coming to cars?* Retrieved from IHS.COM:
   http://blog.ihs.com/is-near-field-communication-nfc-
   finally-coming-to-cars

LLP, A. A. (2017, 1 4). *Near Field Communication Market is
   Expected to Reach $24.0 Billion, Globally by 2020.*
   Retrieved from Open PR:
   http://www.openpr.com/news/405520/Near-Field-
   Communication-Market-is-Expected-to-Reach-24-0-
   Billion-Globally-by-2020.html

Long, K. (2016, 8 21). *UW student project taps ORCA cards,
   unlocks data trove* . Retrieved from seattletimes:
   http://www.seattletimes.com/seattle-
   news/education/uw-student-project-taps-orca-cards-
   unlocks-data-trove/

Lovejoy, B. (2017, 1 11). *iPhone market share grows 6.4% in USA, takes share from Android in most markets*. Retrieved from 9to5mac.com: https://9to5mac.com/2017/01/11/ios-market-share-kantar/

Low, A. (2013, 8 22). *McDonald's NFC Happy Table will be rolling out to rest of Asia*. Retrieved from CNET/Asia: http://asia.cnet.com/mcdonalds-nfc-happy-table-will-be-rolling-out-to-rest-of-asia-62222156.htm?src=twt

Lyman, J. (2017, 3 13). *Why Is the U.S. So Behind on Contactless Payments?* Retrieved from PlugAndPlay: http://plugandplaytechcenter.com/2017/03/13/cont actless-payments-united-states/

MacDailyNews. (2013, 1 22). *Apple iPhone continues lead with 51.2% U.S. market share as Android users increasingly switch to iPhone*. Retrieved from http://macdailynews.com: http://macdailynews.com/2013/01/22/apple-iphone-continues-lead-with-51-2-u-s-market-share-as-android-users-increasingly-switch-to-iphone/

MarketsandMarkets. (2014, 7 31). *Near field communication market worth $16.25 billion by 2022*. Retrieved from http://www.whatech.com/: http://www.whatech.com/market-research-reports/press-release/telecommunications/26025-near-field-communication-market-worth-16-25-billion-by-2022

marketsandmarkets.com. (2015, 11). *Top Market Reports*. Retrieved from MarketsandMarkets:

http://www.marketsandmarkets.com/Market-Reports/near-field-communication-nfc-market-520.html

Mathis, R. (2013, 7 22). *Survey: Why NFC has not fully taken off yet*. Retrieved from SecureIDNews: http://secureidnews.com/news-item/survey-why-nfc-has-not-fully-taken-off-yet/

Mathur, V. (2016, 9 18). *Review: The Arrow Smart Shirt* . Retrieved from Livemint: http://www.livemint.com/Leisure/RTUZItGj1nEBf TdPTYJFEN/Review-The-Arrow-Smart-Shirt.html

McGregor, J. (2013, 8 31). *How high-fiving your front door can let you can ditch the keys*. Retrieved from www.techradar.com: http://www.techradar.com/news/world-of-tech/future-tech/how-high-fiving-your-front-door-can-let-you-can-ditch-the-keys-1175279

Mick, J. (2014, 9 8). *Apple's Lagging Adoption of Token-Protected NFC Mistaken For "Innovation"*. Retrieved from DailyTech: http://www.dailytech.com/Apples+Lagging+Adoptio n+of+TokenProtected+NFC+Mistaken+For+Innova tion/article36518.htm

Microsoft, W. W. (2016, 6 21). *Microsoft Windows Blog*. Retrieved from blog.microsoft.com: https://blogs.windows.com/windowsexperience/201 6/06/21/microsoft-wallet-with-tap-to-pay-is-now-available-for-windows-insiders/#ATQFQrCKqkbOXXet.97

Mobile, S. (2017, 2 13). *Socket Mobile Unveils DuraScan™ D600 Contactless Reader/Writer*. Retrieved from prnewswire: http://www.prnewswire.com/news-releases/socket-mobile-unveils-durascan-d600-contactless-readerwriter-300405851.html

Morris, A. (2017, 9 27). *Swedish rail company swaps paper tickets for embedded microchips*. Retrieved from Dezeen: https://www.dezeen.com/2017/09/27/swedish-rail-company-sj-railways-swaps-train-tickets-for-microchips-news-technology/

Moscow Metro to Introduce Fare Payment Rings. (2017, 10 4). Retrieved from themoscowtimes: https://themoscowtimes.com/news/moscow-metro-to-introduce-payment-ring-59144

Muoio, D. (2017, 10 9). *Digital health news briefs for 10/9/2017*. Retrieved from Mobilehealthnews: http://www.mobihealthnews.com/content/digital-health-news-briefs-1092017

New NFC Forum Specifications Enable NFC Devices to Communicate with Broader Range of Devices and Tags. (2016, 12 19). Retrieved from http://nfc-forum.org: http://nfc-forum.org/newsroom/new-nfc-forum-specifications-enable-nfc-devices-communicate-broader-range-devices-tags/

New NFC Forum Specifications Enable NFC Devices to Communicate with Broader Range of Devices and Tags. (2016, 12 19). Retrieved from http://nfc-forum.org: http://nfc-

forum.org/newsroom/new-nfc-forum-specifications-
enable-nfc-devices-communicate-broader-range-
devices-tags/

News, G. (2016, 2 23). *Gartner Identifies the Top 10 Internet of
Things Technologies for 2017 and 2018.* Retrieved from
gartner.com:
http://www.gartner.com/newsroom/id/3221818

News, N. (2010, 4 20). *Open NFC API for Android™ now
available.* Retrieved from NXP:
http://www.nxp.com/news/press-
releases/2010/04/open-nfc-api-for-android-now-
available.html

Newswire, E. (2013, 12 23). *ABI Research - Smartphones
Accounting for 4 out of 5 NFC Devices as 2013* . Retrieved
from Hispanicbusiness.com:
http://www.hispanicbusiness.com/2013/12/23/abi_r
esearch_-_smartphones_accounting_for.htm

NFC. (2015, 10 14). *NFC Forum Press Releases.* Retrieved from
NFC Forum: http://nfc-forum.org/newsroom/new-
nfc-forum-technical-specifications-broaden-tag-
support-and-enhance-interoperability/

NFC World. (2013, 6 5). *News in brief.* Retrieved from NFC
World:
http://www.nfcworld.com/2013/06/05/324392/dell-
adds-nfc-to-xps-12-laptop/

NFCForum. (2014, 3 20). *NFC Forum.* Retrieved from
http://nfc-forum.org: http://nfc-
forum.org/newsroom/nfc-forum-issues-statement-on-

host-card-emulation/

NFC-Forum. (n.d.). *NFC Forum Technical Specifications*. Retrieved 9 15, 2013, from http://www.nfc-forum.org: http://www.nfc-forum.org/specs/spec_list/

OMAWeb. (n.d.). *About OMA*. Retrieved 9 14, 2013, from http://openmobilealliance.org: http://openmobilealliance.org/about-oma/

Pachal, P. (2016, 1 9). *Mashable Tech*. Retrieved from Mashable: http://mashable.com/2016/01/09/samsung-smart-fashion/#jU5KtiMWAiqB

PhoneArena. (2012, 3 27). *30 million NFC phones were shipped in 2011, huge growth ahead*. Retrieved from www.phonearena.com: http://www.phonearena.com/news/30-million-NFC-phones-were-shipped-in-2011-huge-growth-ahead_id28470

Ransbotham, S. (2017, 1 31). *The Flood of Data From IoT Is Powering New Opportunities — for Some*. Retrieved from http://sloanreview.mit.edu: http://sloanreview.mit.edu/article/the-flood-of-data-from-iot-is-powering-new-opportunities-for-some/

Raschke, K. (2011, February 10). *FareBot: reading ORCA cards on Android*. Retrieved from http://transport.kurtraschke.com: http://transport.kurtraschke.com/2011/02/farebot-orca-android

Research, G. V. (2016, 10). *Near Field Communication (NFC) Market Size Report, 2024* . Retrieved from Grand View Research: http://www.grandviewresearch.com/industry-analysis/near-field-communication-nfc-market

RFIDInsider. (2013, 4 22). *RFID vs. NFC: What's the Difference?* Retrieved from http://blog.atlasrfidstore.com/rfid-vs-nfc: http://blog.atlasrfidstore.com/rfid-vs-nfc#sthash.xqDnmqot.dpuf

Roy, K. (2014, 7 16). *What is Hosted Card Emulation.* Retrieved from www.sequent.com: http://www.sequent.com/host-card-emulation/

Roy, K. (2014). *What is tokenization.* Retrieved from www.sequent.com: http://www.sequent.com/what-is-tokenization/

ScanLife. (2013, 8 1). *Trend Report 2013 Q2.* Retrieved from http://www.scanlife.com: http://www.scanlife.com/trend-reports/q2-2013/CAX346UGKSM

Siemens. (2015 , April). *www.siemens.com/advance.* Retrieved from Siemens.com: https://www.siemens.com/content/dam/mam/tag-siemens-com/projects/customer-magazine/bilder-und-videos/printarchiv/advance/advance-2015-1-en.pdf

simalliance. (2014, 4 29). *simalliance.* Retrieved from http://www.simalliance.org/: http://www.simalliance.org/en/se/se_marketing/sec

ure-element-deployment--host-card-emulation--
v1_hs4nruef.html

Statista. (2014). *Global Apple iPhone Sales since fiscal year 2007*.
Retrieved from The Statistics Portal:
http://www.statista.com/statistics/276306/global-
apple-iphone-sales-since-fiscal-year-2007/

Stemle, C. (2014, 1 31). *BLE vs. NFC: The future of mobile
consumer engagement now [infographic]*. Retrieved from
Mobile Payments Today:
http://www.mobilepaymentstoday.com/blogs/ble-vs-
nfc-the-future-of-mobile-consumer-engagement-now-
infographic/

Tagawa, K. (2016, 6 21). *NFC Forum Blog*. Retrieved from
NFC-forum.org: http://nfc-forum.org/nfc-forum-
chairman-shares-thoughts-hce-looks-feedback/

Tait, D. (2015, 6 29). *Market Insight*. Retrieved from IHS
Markit: https://technology.ihs.com/533599/nfc-
enabled-handset-shipments-to-reach-three-quarters-of-
a-billion-in-2015

Tap2Tag. (2017, 4 24). *PRESS RELEASE - Tap2Tag signs
agreement with Kiroco*. Retrieved from Tap2Tag:
https://www.tap2tag.me/news/press-release-tap2tag-
signs-agreement-with-kiroco/

Tepper, F. (2017, 10 2). *An MLB team is using the iPhone's NFC
feature for contactless stadium entry*. Retrieved from
Techcrunch:
https://techcrunch.com/2017/10/02/an-mlb-team-

is-using-ios-11s-nfc-feature-for-contactless-stadium-entry/

Tepper, F. (2017, 10 2). *An MLB team is using the iPhone's NFC feature for contactless stadium entry*. Retrieved from Techcrunch: https://techcrunch.com/2017/10/02/an-mlb-team-is-using-ios-11s-nfc-feature-for-contactless-stadium-entry/

Thinfilm's tiny printed NFC labels enable brands to connect with consumers via smart packaging. (2017, 3 23). Retrieved from venturebeat: https://venturebeat.com/2017/03/23/thinfilms-tiny-printed-nfc-labels-enable-brands-to-connect-with-consumers-via-smart-packaging/

Vanderhoof, R. (2013). Mobile/NFC Security Fundamentals. *Smart Card Alliance and Global Platform Webinar.* http://www.smartcardalliance.org/resources/webinars/Anatomy_of_a_Mobile_Device_030513.pdf. Retrieved from http://www.smartcardalliance.org/resources/webinars/Anatomy_of_a_Mobile_Device_030513.pdf

Whitney, L. (2014, 9 16). *Subway to use Softcard NFC for mobile payments*. Retrieved from www.cnet.com: http://www.cnet.com/news/subway-restaurants-to-use-softcard-nfc-for-mobile-payments/

wiiudaily. (2014). *http://wiiudaily.com/wii-u-nfc/*. Retrieved from http://wiiudaily.com: http://wiiudaily.com/wii-u-nfc/

XDA. (2013, 8 20). *NFC tags*. Retrieved from XDA Developer

Forum: http://forum.xda-
developers.com/wiki/NFC_Tags

Your Automobile is Your Biggest Connected Mobile Device. (2016, 11
10). Retrieved from NFC Forum: http://nfc-
forum.org/automobile-biggest-connected-mobile-
device/

About the Author

Hsuan-hua Chang is a consultant, coach, writer and speaker.

She has over 20 years of experience in Wireless technology as a technical product manager, marketing evangelist, architect and developer. Her focus is on NFC, IoT and Cloud Computing.

Hsuan-hua has been coaching executives and entrepreneurs since 2004 as a certified leadership coach. She empowers business leaders through the use of technology and leadership development.

She holds a Master's Degree in Computer Science and a Leadership MBA. The combination of her enterprise technology experience, with professional business coaching, results in a unique perspective and approach that explores business opportunities in technology. Her goal is to leverage technologies and develop leaders in order to make our world a better place.

Learn more about Hsuan-hua Chang at:
www.linkedin.com/in/hsuanhuachang and
www.coachseattle.com

Made in the USA
Columbia, SC
22 July 2020